Collins

The Shanghai Maths Project

For the English National Curriculum

Practice Book 1B

Series Editor: Professor Lianghuo Fan

UK Curriculum Consultant: Paul Broadbent

PUBLISHERS

Since 1817

William Collins' dream of knowledge for all began with the publication of his first book in 1819.

A self-educated mill worker, he not only enriched millions of lives, but also founded a flourishing publishing house. Today, staying true to this spirit, Collins books are packed with inspiration, innovation and practical expertise. They place you at the centre of a world of possibility and give you exactly what you need to explore it.

Collins. Freedom to teach.

Published by Collins, an imprint of HarperCollins*Publishers*
The News Building
1 London Bridge Street
London
SE1 9GF

HarperCollins *Publishers*
Macken House,
39/40 Mayor Street Upper,
Dublin 1,
D01 C9W8
Ireland

Browse the complete Collins catalogue at
www.collins.co.uk

10 9 8 7

978-0-00-822608-4

Translated by Professor Lianghuo Fan, Adapted by Professor Lianghuo Fan.

British Library Cataloguing in Publication Data

A catalogue record for this publication is available from the British Library.

Series Editor: Professor Lianghuo Fan
UK Curriculum Consultant: Paul Broadbent
Publishing Manager: Fiona McGlade
In-house Editor: Nina Smith
In-house Editorial Assistant: August Stevens
Project Manager: Emily Hooton
Copy Editor: Tracy Thomas
Proofreaders: Dawn Booth and Tracy Thomas
Cover design: Kevin Robbins and East China Normal University Press Ltd
Cover artwork: Daniela Geremia
Internal design: 2Hoots Publishing Services Ltd
Typesetting: 2Hoots Publishing Services Ltd
Illustrations: QBS
Production: Rachel Weaver
Printed and bound in the UK using 100% Renewable Electricity at CPI Group (UK) Ltd

The Shanghai Maths Project (for the English National Curriculum) is a collaborative effort between HarperCollins, East China Normal University Press Ltd. and Professor Lianghuo Fan and his team. Based on the latest edition of the award-winning series of learning resource books, *One Lesson, One Exercise*, by East China Normal University Press Ltd. in Chinese, the series of Practice Books is published by HarperCollins after adaptation following the English National Curriculum.

Practice Book Year 1B is translated and developed by Professor Lianghuo Fan with assistance of Ellen Chen, Ming Ni, Huiping Xu and Dr. Lionel Pereira-Mendoza, with Paul Broadbent as UK Curriculum Consultant.

MIX
Paper | Supporting responsible forestry
FSC™ C007454

This book is produced from independently certified FSC™ paper to ensure responsible forest management.

For more information visit: www.harpercollins.co.uk/green

Contents

Chapter 7 Introduction to time (I)

Chapter 8 Let's practise geometry

Chapter 4 Recognising shapes

4.1 Shapes of objects (1)

 Learning objective Recognise and name 3-D shapes

 Basic questions

1 Draw lines to match the solid figures with the shapes below.

2 Count the number of each shape and then fill in the boxes.

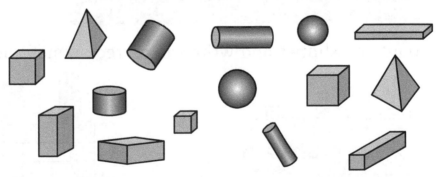

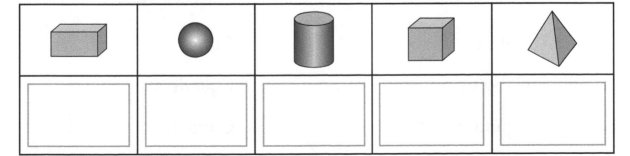

3 Try rolling each shape and then fill in the table with a tick ✓ or a cross ✗. One has been done for you.

	Cannot roll	Can roll	
		in one direction	in all directions
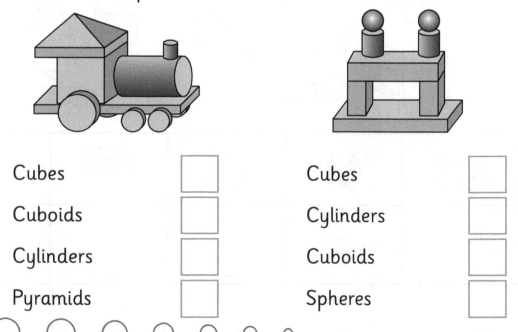	✓		

Challenge and extension question

4 Count the shapes and write the correct number in the box.

Cubes ☐   Cubes ☐

Cuboids ☐   Cylinders ☐

Cylinders ☐   Cuboids ☐

Pyramids ☐   Spheres ☐

4.2 Shapes of objects (2)

 Learning objective Recognise and name 3-D shapes

 Basic questions

1 Draw lines to match the objects with the shapes. One has been done for you.

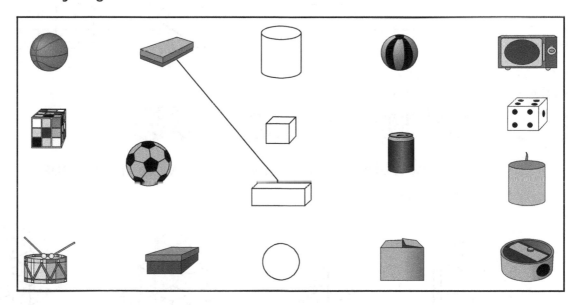

2 Count the shapes and write the correct number in the box.

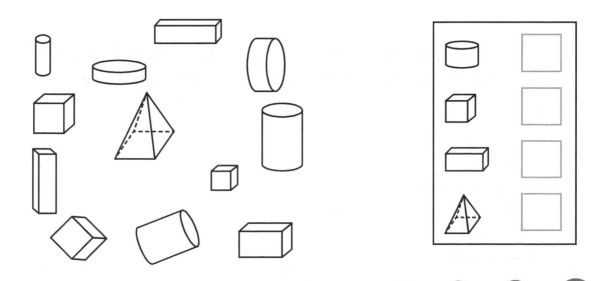

3 Draw lines to match each object to the correct shape below.

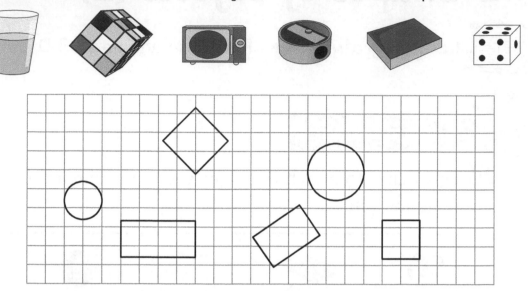

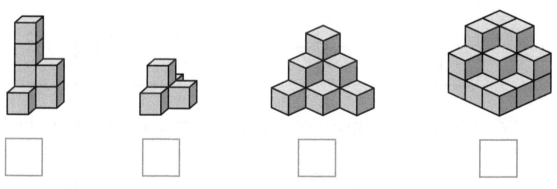

Challenge and extension questions

4 Count the cubes in each tower and write the number underneath.

5 3 triangles are formed using 9 sticks. Move 3 of the sticks to form 5 triangles. Draw how you did it in the box.

Chapter 4 test

1 Calculate the following mentally.

$5 + 2 = \boxed{}$ $\qquad$ $3 + 7 = \boxed{}$ $\qquad$ $14 - 6 = \boxed{}$

$\boxed{} + 6 = 10$ $\qquad$ $4 + 2 = \boxed{}$ $\qquad$ $18 - 6 = \boxed{}$

$17 - 3 = \boxed{}$ $\qquad$ $15 = \boxed{} + 4$ $\qquad$ $9 - 3 = \boxed{}$

$4 + 6 = \boxed{}$ $\qquad$ $9 + 8 = \boxed{}$ $\qquad$ $6 = \boxed{} - 2$

$15 + 2 = \boxed{}$ $\qquad$ $10 + 7 = \boxed{}$ $\qquad$ $14 + 4 = \boxed{}$

$\boxed{} - 3 = 9$ $\qquad$ $7 - 3 = \boxed{}$ $\qquad$ $5 + 7 = \boxed{}$

$7 + 7 = \boxed{}$ $\qquad$ $7 + \boxed{} = 13 - \boxed{}$

2 Count and then compare and add.

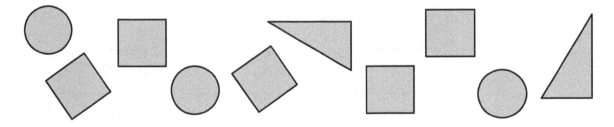

Shape	How many?
$\square$	
$\bigcirc$	
◺	

There are ____ more $\boxed{}$ than $\bigcirc$.

There are ____ fewer ◺ than $\boxed{}$.

There are ____ $\boxed{}$ $\bigcirc$ and ◺ altogether.

$\boxed{} + \boxed{} + \boxed{} = \boxed{}$.

3 Count the shapes and then write the correct number in each box.

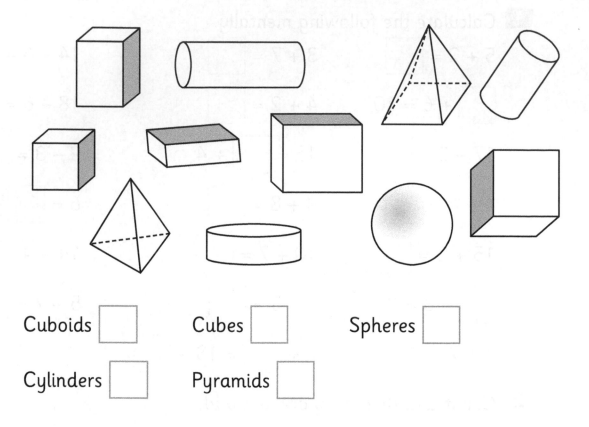

Cuboids ☐   Cubes ☐   Spheres ☐

Cylinders ☐   Pyramids ☐

4 Finish the number walls.

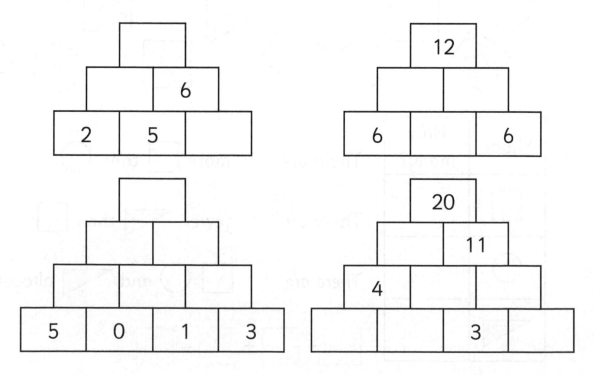

5 Write the correct numbers for each shape below.

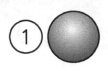 ① 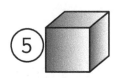 ② 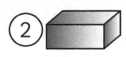 ③ 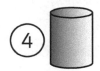 ④

⑤ ⑥ ⑦ 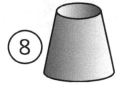 ⑧

(a) These shapes have rectangular faces ▭ _____

(b) These shapes have square faces ☐ _____

(c) These shapes have triangular faces △ _____

(d) These shapes have circular faces ○ _____

6 Write the number sentences for these pictures.

9 + ☐ = ☐   ☐ + ☐ = ☐

15 − ☐ = ☐   ☐ − ☐ = ☐

7 Look carefully and then complete each sentence.

(a) ▭ There are ☐ rectangles altogether.

(b) There are ☐ cubes altogether.

Chapter 5 Consolidation and enhancement

5.1 Sorting shapes

 Learning objective Recognise and sort 2-D shapes

 Basic questions

1 Count and then write the correct number in each box.

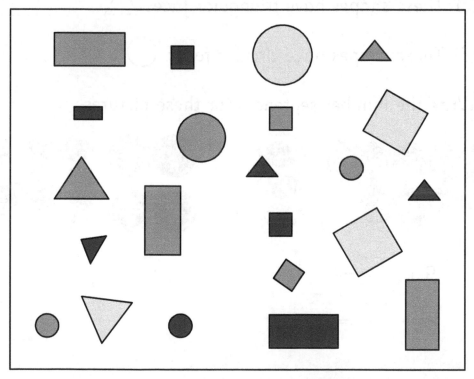

(a) Sort by shape.

(b) Sort by size.　Big　Small

2 Sort the shapes in different ways and then put the numbers into the ovals.

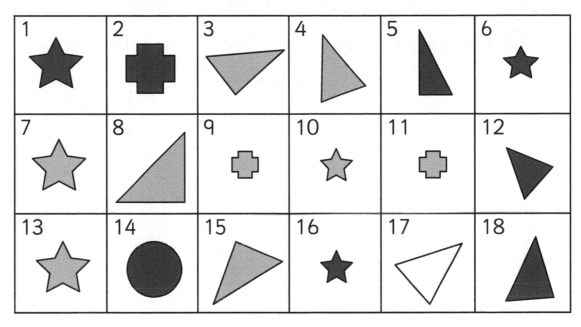

(a) Sort by _____ .

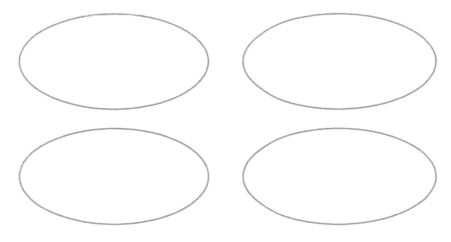

(b) Sort by _____ .

Challenge and extension question

3 Complete each statement. One has been done for you.

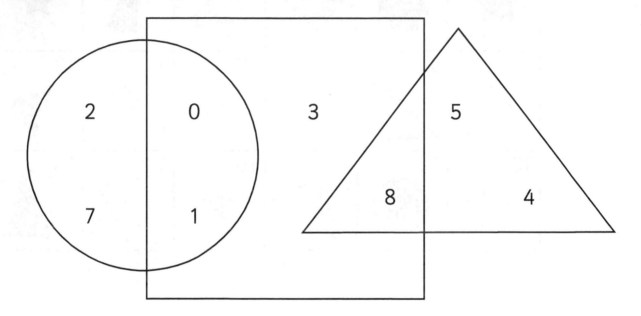

The numbers in the ⭕ are 0, 1, 2, 7 _____.

The numbers in the ⬜ are _____.

The numbers in both ⭕ and ⬜ are _____.

The number in both △ and ⬜ is _____.

5.2 Calculating with reasoning

 Learning objective Add and subtract numbers to 20

 Basic questions

1 Complete the addition and subtraction problems.

8 + 4 = ☐   8 + 5 = ☐   8 + ☐ = ☐

12 − 8 = ☐   ☐ − 8 = ☐   14 − ☐ = ☐

12 − 4 = ☐   13 − ☐ = ☐   ☐ − ☐ = ☐

2 Write the correct answers on the floors of the buildings.

+ 6	
10	
11	
12	
13	
14	

+ 7	
10	
9	
8	

8 +	
4	
5	
6	

− 4	
16	
15	
14	
13	
12	

− 6	
11	
12	
13	

15 −	
9	
8	
7	

3 Complete these calculations.

8 + 4 = ☐   18 − 10 = ☐   15 − 7 = ☐

8 + 5 = ☐   10 + 8 = ☐   14 − 7 = ☐

9 + 5 = ☐   11 + 8 = ☐   7 + 5 = ☐

18 − 8 = ☐   11 + 9 = ☐   8 + 6 = ☐

18 − 9 = ☐   16 − 7 = ☐   9 + 7 = ☐

4 Fill in the boxes with suitable numbers. (Think about the relationships between the numbers in the number sentences.)

5 + 2 = ☐   8 − 6 = ☐

15 + 2 = ☐   18 − 6 = ☐

25 + ☐ = ☐   28 − ☐ = ☐

35 + ☐ = ☐   38 − ☐ = ☐

Challenge and extension question

5 Can you work out the number that each symbol stands for?

△ ○ □   19

△ △ △   15

♡ ☺ △   14

12   20   16

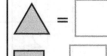
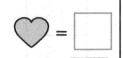
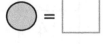

△ = ☐   ♡ = ☐

□ = ☐   ○ = ☐

☺ = ☐

5.3 Comparing numbers

 Learning objective Compare numbers to 20

 Basic questions

1 Read carefully and then draw.

(a) Draw as many △ as ☆.

(b) Draw 1 more △ than ◇.

◇ ◇ ◇

(c) Draw 1 fewer △ than ◯.

◯ ◯ ◯ ◯

(d) Draw 2 more △ than ♡.

♡ ♡ ♡ ♡

(e) Draw 3 fewer △ than ☺.

☺ ☺ ☺ ☺ ☺ ☺

2 Look carefully and then write >, < or = in each ( ).

(a) 6 ( ) 8   12 ( ) 9   15 ( ) 20

  18 ( ) 8   16 ( ) 14   1 ( ) 0

(b) 9 + 6 ( ) 17   16 − 5 ( ) 9   15 ( ) 20 − 5

  12 + 8 ( ) 18   15 − 6 ( ) 11   17 ( ) 7 + 7

(c) 7 + 9 ( ) 19 − 2   18 − 3 ( ) 18 − 2   20 − 6 ( ) 8 + 8

  8 + 7 ( ) 6 + 9   20 − 4 ( ) 11 + 5   5 + 9 ( ) 17 − 5

3 Fill in the boxes with suitable numbers.

(a) 18 > ☐   10 < ☐   7 > ☐

  5 < ☐ < ☐ < 10   18 > ☐ > 16 > ☐

(b) 8 + ☐ < 14   ☐ − 10 < 3   ☐ > 16 − 5

  20 − ☐ > 10   ☐ + 4 > 12   11 < ☐ + 6

(c) ☐ + 4 < 20 − 4   ☐ − 6 > 4 + 4   10 − 7 > ☐ − 8

  16 − ☐ < 7 + 7   3 + 8 < ☐ − 5   6 + 7 < 9 + ☐

Challenge and extension questions

4 Without calculating, write >, < or = in each $\bigcirc$.

24 + 6 $\bigcirc$ 24 + 8   55 + 9 $\bigcirc$ 54 + 9

35 + 7 $\bigcirc$ 33 + 9   48 − 5 $\bigcirc$ 48 − 8

66 − 9 $\bigcirc$ 56 − 9   55 − 7 $\bigcirc$ 54 − 6

5 Work out the largest number you can put in each box.

7 + ☐ < 15   18 > ☐ + 9

8 + ☐ < 20 − 8   ☐ − 6 < 5

7 < 13 − ☐   20 − ☐ > 6 + 6

5.4 Half and quarter

 Learning objective Recognise a half and a quarter of shapes and quantities

 Basic questions

1 Has each shape been halved? (Put a ✔ for yes and a ✘ for no.)

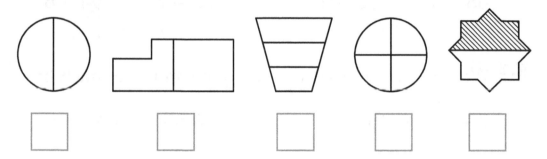

2 Write **half** or **quarter** to show the shaded part of each shape.

(a)

_____

(b)

_____

(c)

_____

(d)

_____

3 Follow the instructions to colour each shape.

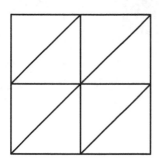 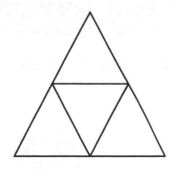 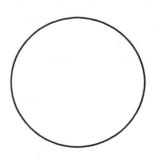

Colour 1 half   Colour 1 quarter   Colour 1 whole

4 Halve each group of dots and then write the number sentences.

○ ○ ○   ○ ○ ○ ○ ○ ○ ○   ○ ○ ○ ○ ○
‾‾‾‾‾‾‾‾
○ ○ ○   ○ ○ ○ ○ ○ ○ ○   ○ ○ ○ ○ ○

☐ = ☐ + ☐   ☐ = ☐ + ☐   ☐ = ☐ + ☐

5 Write the correct numbers in the circles.

2 $\xrightarrow{\text{double}}$ ◯   ◯ $\xrightarrow{\text{double}}$ 6

5 $\xrightarrow{\text{double}}$ ◯ $\xrightarrow{\text{double}}$ ◯

8 $\xrightarrow{\text{half}}$ ◯   ◯ $\xrightarrow{\text{half}}$ 8

12 $\xrightarrow{\text{half}}$ ◯ $\xrightarrow{\text{half}}$ ◯

6 What number does each shape stand for?

△ + △ = 10   ◯ + ◯ = 18   ☆ + ☆ = 14

△ = ☐   ◯ = ☐   ☆ = ☐

Challenge and extension questions

7 Think carefully and then fill in the ☐.

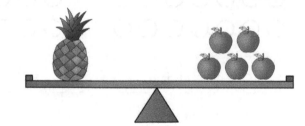

8 Samira had 20 marbles. She gave half of them to Ella and a quarter of them to Finn.

(a) How many marbles did Ella receive? ☐

(b) How many marbles did Finn receive? ☐

(c) How many marbles did Samira still have? ☐

5.5 Let's do additions together

Learning objective Use number bonds to 20

Basic questions

1 Complete the addition table.

0 + 0	1 + 0	2 + 0	3 + 0	4 + 0		6 + 0			9 + 0	
	1 + 1	2 + 1			5 + 1		7 + 1			10 + 1
0 + 2		2 + 2	3 + 2	4 + 2		6 + 2		8 + 2	9 + 2	
	1 + 3		3 + 3		5 + 3		7 + 3			10 + 3
0 + 4		2 + 4		4 + 4		6 + 4		8 + 4	9 + 4	
	1 + 5	2 + 5		4 + 5	5 + 5		7 + 5		9 + 5	
	1 + 6		3 + 6		5 + 6	6 + 6		8 + 6		10 + 6
0 + 7	1 + 7		3 + 7			6 + 7	7 + 7	8 + 7		10 + 7
	1 + 8			4 + 8				8 + 8	9 + 8	
0 + 9	1 + 9	2 + 9		4 + 9		6 + 9			9 + 9	10 + 9
0 + 10			3 + 10	4 + 10			7 + 10	8 + 10		10 + 10

2 Colour the addition calculations with the answers of

(4), (7), (10), (13) and (16), in the above table.

3 Use the grid from question 1 to help you to complete these sections.

		9 + 7 =
	8 + 8 =	
7 + 9 =		

5 + 7 =		
	4 + 8 =	
		3 + 9 =

4 Write the numbers on each floor of the towers.

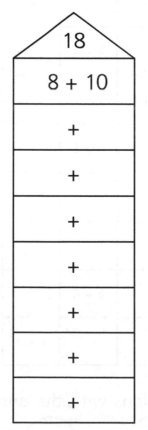

13
3 + 10
4 + 9
5 + 8
+
+
+
+
+

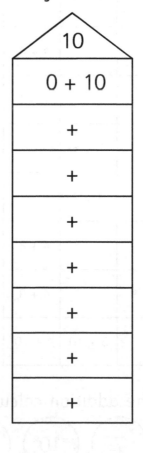

18
8 + 10
+
+
+
+
+
+
+

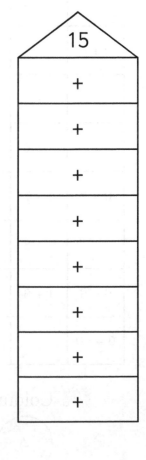

10
0 + 10
+
+
+
+
+
+
+

15
+
+
+
+
+
+
+
+

5 Make up addition problems and write the number sentences.

(a) There were 10 birds in the tree. Another 8 birds joined them.

_____?

Number sentence: □ ○ □ ○ □

(b) There are 12 girls and 8 boys in the playground.

_____?

Number sentence: □ ○ □ ○ □

(c) After 9 cars drove away, there were 6 cars left in the car park.

_____?

Number sentence: □ ○ □ ○ □

Challenge and extension question

6 Fill in the ○ with

numbers ① , ② ,

③ , ④ , ⑤ ,

⑥ , ⑦ so that the

sum of the 3 numbers on each line is 16.

5.6 Let's do subtractions together

 Learning objective Use number bonds to subtract numbers within 20

 Basic questions

1 Complete the subtraction table.

0 – 0	1 – 0	2 – 0		4 – 0					9 – 0	
	2 – 1	3 – 1			6 – 1		8 – 1			11 – 1
2 – 2		4 – 2	5 – 2	6 – 2		8 – 2		10 – 2		
				7 – 3						13 – 3
4 – 4		6 – 4			9 – 4					
		7 – 5	8 – 5					13 – 5		
6 – 6	7 – 6		9 – 6			12 – 6				16 – 6
	8 – 7					13 – 7				17 – 7
	9 – 8	10 – 8								
9 – 9	10 – 9	11 – 9			14 – 9			17 – 9		19 – 9
10 – 10	11 – 10			14 – 10			17 – 10	18 – 10		20 – 10

2 Colour the subtraction calculations with the answers of (0), (3), (6) and (9) in the above table, using one colour.

3 Use the grid from question 1 to help you to complete these sections.

		8 – 3 =
	8 – 4 =	
8 – 5 =		

		12 – 3 =
	12 – 4 =	
12 – 5 =		

4 Write the numbers on each floor of the towers.

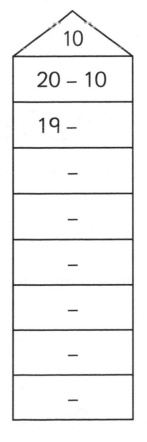

12
13 – 1
14 – 2
15 – 3
–
–
–
–
–

8
10 – 2
11 –
–
–
–
–
–
–

10
20 – 10
19 –
–
–
–
–
–
–

5
10 – 5
11 –
–
–
–
–
–
–

5 Make up a subtraction problem for each question.

(a) There were 20 cherries on the plate. Yaseen ate 8 of them.

_____?

Number sentence: ☐ ◯ ☐ ◯ ☐

(b) There are 15 white rabbits and black rabbits in the field. 8 of them are white rabbits.

_____?

Number sentence: ☐ ◯ ☐ ◯ ☐

(c) There are 12 red beanbags and 9 green beanbags in the crate.

_____?

Number sentence: ☐ ◯ ☐ ◯ ☐

Challenge and extension question

6 Make matching number sentences using the numbers in the circles.

⑤ ⑥ ⑦ ⑧ ⑨ ⑩ ⑪ ⑫

You can only use each number once in each sentence.

(a) ☐ – ☐ = ☐ – ☐ = ☐ – ☐ = ☐ – ☐ .

(b) ☐ – ☐ = ☐ – ☐ = ☐ – ☐ = ☐ – ☐ .

5.7 Making number sentences

 Basic questions

1 Look at each picture and write the number sentences.

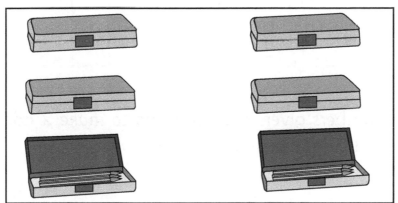

☐ + ☐ = ☐   ☐ ◯ ☐ = ☐

☐ + ☐ = ☐   ☐ ◯ ☐ = ☐

☐ + ☐ = ☐   ☐ ◯ ☐ = ☐

☐ + ☐ = ☐   ☐ ◯ ☐ = ☐

2 Work out the answer for each calculation.
Underneath, write different number sentences
for the same total.

9 + 6 = ☐

☐ + ☐ = ☐

☐ + ☐ = ☐

☐ + ☐ = ☐

12 + 8 = ☐

☐ + ☐ = ☐

☐ + ☐ = ☐

☐ + ☐ = ☐

16 – 9 = ☐

☐ + ☐ = ☐

☐ + ☐ = ☐

☐ – ☐ = ☐

20 – 6 = ☐

☐ + ☐ = ☐

☐ + ☐ = ☐

☐ – ☐ = ☐

3 Use the 3 numbers given in each group to make 2 addition
sentences and 2 subtraction sentences.

7 8 15   5 20 15   9 9 0

_____   _____   _____

_____   _____   _____

_____   _____   _____

_____   _____   _____

4 Choose 3 numbers in each group and make 2 addition sentences and 2 subtraction sentences.

    4   6   8   12   5   9   12   14

_____

_____

_____

_____

Challenge and extension question

5 Put the 9 numbers below into 3 groups. Then use the 3 numbers in each group to make 2 addition sentences and 2 subtraction sentences.

2, 6, 7, 8, 9, 10, 12, 14, 16

5.8 Mathematics playground (1)

 Learning objective Read, write, add and subtract numbers to 20

 Basic questions

1 Calculate mentally. Write the answers in the boxes.

3 + 5 = ☐   10 – 7 = ☐   6 + 4 = ☐   5 + 4 + 6 = ☐

9 + 0 = ☐   8 – 6 = ☐   7 + 8 = ☐   8 + 9 – 5 = ☐

13 + 2 = ☐   19 – 9 = ☐   11 – 6 = ☐   18 – 5 – 7 = ☐

7 + 7 = ☐   14 – 5 = ☐   17 – 0 = ☐   16 – 6 + 5 = ☐

8 – 8 = ☐   12 – 3 = ☐   9 + 9 = ☐   6 + 6 + 6 = ☐

2 Fill in the boxes with suitable numbers.

(a)

8			12		15				

	2	4		10		14			

19	17				7			

(b) 6 ones and 1 ten make ☐. 2 tens make ☐.

There are ☐ ones and ☐ ten in 18.

There are ☐ fives in 20.

(c)

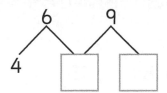

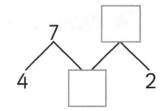

(d) Write the numbers before or after.

	8	

12		14

	19	

3 Look at the pictures and write the number sentences.

(a)

☐ + ☐ = ☐

☐ + ☐ = ☐

☐ − ☐ = ☐

☐ + ☐ = ☐

(b)

☐ ◯ ☐ ◯ ☐ = ☐

(c)

☐ ◯ ☐ ◯ ☐ = ☐

4 Write >, < or = in each ◯.

7 ◯ 9   8 + 8 ◯ 15

12 ◯ 20 – 6   7 + 4 ◯ 19 – 8

8 ◯ 12   12 – 7 ◯ 6

11 ◯ 6 + 6   12 + 5 ◯ 14 + 5

18 ◯ 16   13 – 8 ◯ 5

20 ◯ 5 + 15   18 – 8 ◯ 16 – 5

Challenge and extension question

5 Write > or < in each ◯.

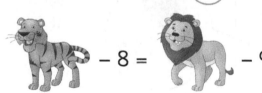 – 8 = – 9

 ◯

 – 10 =  + 10

 ◯

5.9 Mathematics playground (2)

 Learning objective Add and subtract numbers to 20

 Basic questions

1 Calculate mentally. Write the answers in the boxes.

$9 + 9 = \boxed{}$ $5 + 5 = \boxed{}$ $6 + 14 = \boxed{}$ $8 + 4 + 6 = \boxed{}$

$16 - 7 = \boxed{}$ $9 - 6 = \boxed{}$ $17 - 10 = \boxed{}$ $7 + 9 - 5 = \boxed{}$

$16 - 8 = \boxed{}$ $8 + 7 = \boxed{}$ $11 - 6 = \boxed{}$ $18 - 10 - 8 = \boxed{}$

2 Fill in the $\boxed{}$ with the correct numbers.

$5 + \boxed{} = 11$ $12 - \boxed{} = 8$ $8 + 2 + \boxed{} = 16$

$\boxed{} + 10 = 19$ $16 - \boxed{} = 5$ $20 - 7 - \boxed{} = 0$

$12 + \boxed{} = 20$ $\boxed{} - 10 = 7$ $6 + \boxed{} - 8 = 2$

3 Use the 3 numbers given in each group to make 2 addition sentences and 2 subtraction sentences.

4   8   12

3   13   10

8   8   0

4 Write the correct numbers in the table.

Double it								Halve it
	10		4		3		8	
12		18		10		14		

5 Put the calculations in order from the smallest to the largest.

16 – 10   14 + 3   9 + 9

17 – 12   11 + 4

6 Look at the shape of each object. Then write the number of each object into the correct oval.

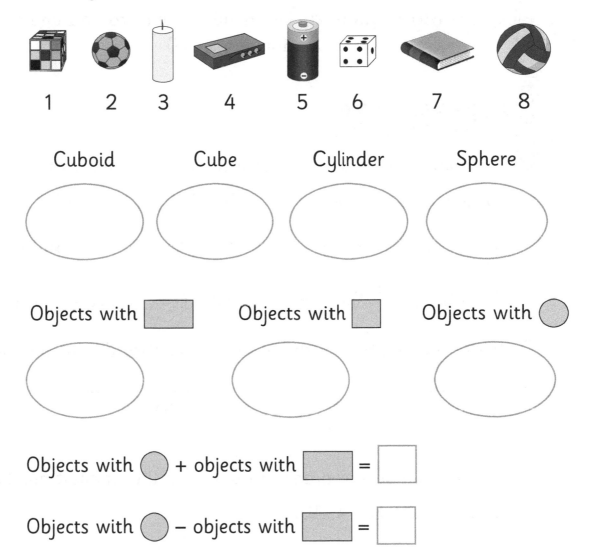

1   2   3   4   5   6   7   8

Cuboid   Cube   Cylinder   Sphere

Objects with ▭   Objects with ▭   Objects with ◯

Objects with ◯ + objects with ▭ = ☐

Objects with ◯ − objects with ▭ = ☐

Challenge and extension question

7 Continue the pattern and draw 11 more beads onto the string.

5.10 Mathematics playground (3)

 Learning objective Use number bonds to add and subtract numbers to 20

 Basic questions

1 Calculate mentally. Write your answers in the boxes.

$7 + 7 =$ ☐ $9 + 6 =$ ☐ $5 + 13 =$ ☐ $7 + 5 + 5 =$ ☐

$12 - 7 =$ ☐ $18 - 6 =$ ☐ $18 - 10 =$ ☐ $4 + 8 - 5 =$ ☐

$16 - 9 =$ ☐ $9 + 7 =$ ☐ $12 - 6 =$ ☐ $16 - 1 - 8 =$ ☐

2 Write the numbers in the boxes to complete the number line.

☐   2   ☐   ☐   ☐   10   ☐   ☐   ☐

3 Put the numbers in order, starting with the smallest number.

5, 12, 0, 20, 8, 14, 10, 3

4 Fill in the numbers on each floor of the towers.

(12)	(20)	(8)	(5)
12 + 0	10 + 10	18 – 10	10 – 5
11 + 1	11 + 9	17 – 9	11 – 6
+	+	–	–
+	+	–	–
+	+	–	–

5 Complete the number walls.

(a)
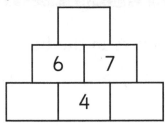

(b)
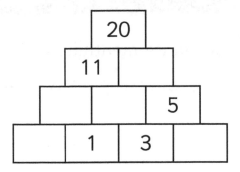

6 Write >, < or = in each ◯ and a number in each ☐.

5 + 9 ◯ 15   12 − 4 ◯ 7   4 + 6 ◯ 7 + 3

16 − 5 ◯ 11 − 5   9 > 2 + ☐   ☐ + 7 < 15

7 > ☐ + 4   ☐ − 5 = 7 + ☐   10 − ☐ = ☐ − 6

Challenge and extension question

7 Use these numbers to fill in the ◯, so that the total of the 3 numbers on both lines are the same.

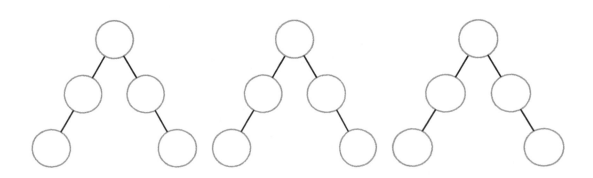

35

Chapter 5 test

1 Calculate mentally. Write the answers in the boxes.

9 + 4 = ☐   15 − 6 = ☐   5 + 7 = ☐   14 − 8 = ☐

7 + 4 = ☐   17 − 9 = ☐   15 − 7 = ☐   10 + 8 = ☐

14 − 8 = ☐   15 − 8 = ☐   9 + 7 = ☐   14 − 9 = ☐

7 + 8 = ☐   9 + 3 = ☐   13 − 4 = ☐   6 + 7 = ☐

2 Use the pictures to write addition and subtraction number sentences.

☐ + ☐ = ☐

☐ + ☐ = ☐

☐ − ☐ = ☐

☐ − ☐ = ☐

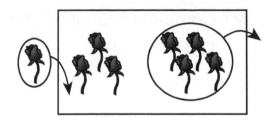

There are 16  in total.

If two are out of the basket,
how many  are in the basket?

☐ ◯ ☐ = ☐   ☐ ◯ ☐ ◯ ☐ = ☐

3 Write the answer in each box.

(a) 1 ten and 5 ones make ☐.

(b) There is ☐ ten and ☐ ones in 17.

(c) The sum of the greatest 1-digit number and the smallest
2-digit number is ☐.

(d) Put the numbers 18, 4, 15, 2, 20 and 0 in order starting
from the largest.

_____

4 Complete the number patterns.

11	12			15		

	7		3

10		6		2

		14	

5 Write the correct numbers in the blanks.

6 $\xrightarrow{\text{double}}$ _____

8 $\xrightarrow{\text{half}}$ _____ $\xrightarrow{\text{half}}$ _____

20 $\xrightarrow{\text{half}}$ _____ $\xrightarrow{\text{half}}$ _____

7 $\xrightarrow{\text{double}}$ _____

6 Complete the calculations and then write your own calculations.

3 + 11 = ☐   20 − 3 = ☐   17 − 8 = ☐

4 + 10 = ☐   18 − 5 = ☐   15 − 8 = ☐

5 + ☐ = ☐   16 − ☐ = ☐   13 − ☐ = ☐

☐ + ☐ = ☐   ☐ − ☐ = ☐   ☐ − ☐ = ☐

7 Complete the number walls.

(a)

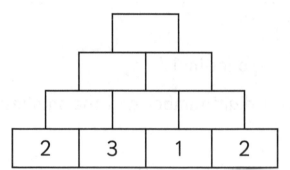

| 2 | 3 | 1 | 2 |

(b)

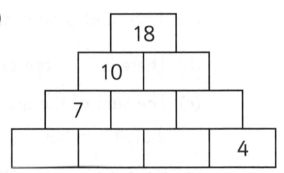

8 Write >, < or = in each ◯ and a number in each ☐ .

7 + 8 ◯ 15   12 − 3 ◯ 8   7 + 6 ◯ 8 + 5

16 − 5 ◯ 15 − 6   9 > 4 + ☐   20 − ☐ < 5

☐ + 6 < 12   17 > ☐ + 10   ☐ + 7 = ☐ − 5

12 − ☐ = ☐ − 9   14 − ☐ = 7 + ☐

9 Complete addition and subtraction tables.

3 + 3 =			9 − 4 =		11 − 4 =
3 + 4 =	4 + 4 =			11 − 5 =	
	4 + 5 =				13 − 6 =

10 Write the number sentences and calculate.

(a) Minna has 8 storybooks and 5 science books. How many books does she have in total?

(b) Levi had 15 marbles. He gave away 7 of them. How many marbles did he still have?

(c) 14 pupils in Class 1 learned how to swim. 9 of them are boys. How many are girls?

(d) 7 brown ducks were on the pond. Some white ducks joined them later. Now there are 12 ducks altogether. How many white ducks joined in?

☐ ○ ☐ ○ ☐

(e) After 8 cars drove away from the car park, there were 12 cars left. How many cars were there at first?

☐ ○ ☐ ○ ☐

(f) There were 15 birds in the tree. 9 birds flew away. Then another 10 birds flew in to the tree. How many birds are in the tree now?

☐ ○ ☐ ○ ☐ ○ ☐

11 Colour the shape that does not belong to the group.

6.1 Tens and ones

Learning objective Partition numbers to 100 into tens and ones

Basic questions

1 Count and then fill in the missing numbers.

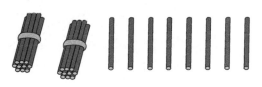

☐ tens and ☐ ones

make ☐

☐ + ☐ = ☐

☐ tens and ☐ ones

make ☐

☐ + ☐ = ☐

☐ tens and ☐ ones

make ☐

☐ + ☐ = ☐

☐ tens and ☐ ones

make ☐

☐ + ☐ = ☐

2 Group the objects in tens. Then count in tens and ones, record the result and write the numbers in numerals.

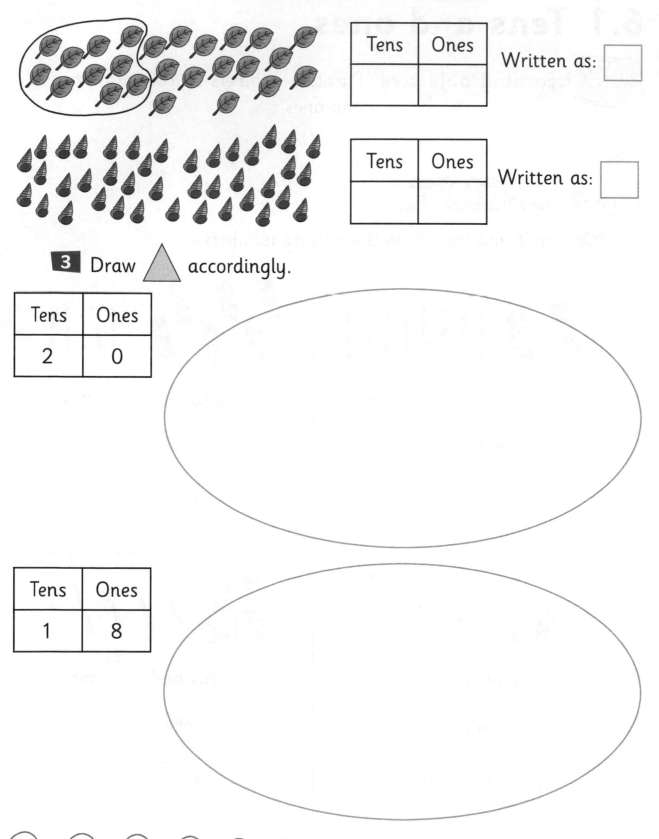

Tens	Ones

Written as: ☐

Tens	Ones

Written as: ☐

3 Draw △ accordingly.

Tens	Ones
2	0

Tens	Ones
1	8

4 Fill in the ☐ with suitable numbers.

(a) 50 is ☐ tens.   3 tens are ☐.   80 is ☐ tens.

4 tens are ☐.   100 is ☐ tens.   ☐ ones are 100.

(b) 26 = 20 + ☐   53 = ☐ + 3   84 = 80 + ☐

69 = ☐ + 9   71 = 70 + ☐   45 = ☐ + ☐

Challenge and extension question

5 Fill in the boxes.

(a) 5 tens and 2 ones make ☐

4 ones and 9 tens make ☐.

(b) 63 is made up of ☐ tens and ☐ ones.

There are ☐ tens and ☐ ones in 78.

(c) In the number 36, 3 is in the _____ place.

It means _____ _____.

6 is in the _____ place.

It means _____ _____.

6.2 Knowing 100

Learning objective Add tens to total 100 and subtract tens from 100

Basic questions

1 Complete the number lines.

0　10　20　□　□　□　60　□　□　□　□

0　10　□　30　□　50　□　70　□　90　□

2 Write numbers in the boxes using the 100 square diagram below.

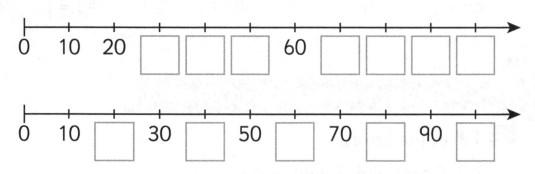

3 tens are □.

5 tens are □.

8 tens are □.

There are □ groups of 10 in 100.

There are □ groups of 50 in 100.

There are □ groups of 25 in 100.

3 Use the 100 square diagram in Question 2 to work out the answers.

10 + 40 = ☐	100 − 60 = ☐	90 − 50 = ☐
20 + 40 = ☐	90 − 50 = ☐	80 − 50 = ☐
30 + 40 = ☐	80 − 40 = ☐	70 − 50 = ☐
40 + 40 = ☐	70 − 30 = ☐	60 − 50 = ☐
50 + 40 = ☐	60 − 20 = ☐	50 − 50 = ☐

10 + 90 = ☐	90 + ☐ = 100	100 − ☐ = 50
20 + 70 = ☐	80 + ☐ = 100	100 − ☐ = 40
30 + 50 = ☐	70 + ☐ = 100	100 − ☐ = 30
40 + 30 = ☐	60 + ☐ = 100	100 − ☐ = 20
50 + 10 = ☐	50 + ☐ = 100	100 − ☐ = 10

4 Work out the missing numbers using the 100 square diagram.

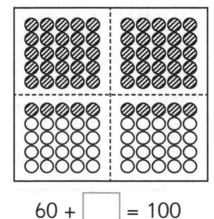
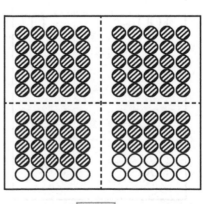

60 + ☐ = 100   85 + ☐ = 100

45

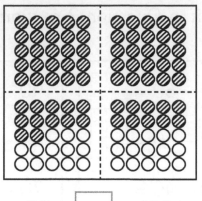
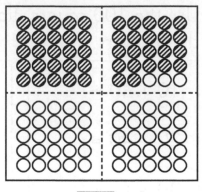

72 + ☐ = 100   47 + ☐ = 100

Challenge and extension question

5 Use a pattern to complete the number sentences.

(a)

10 + ☐ = 100	92 + ☐ = 100
15 + ☐ = 100	82 + ☐ = 100
20 + ☐ = 100	72 + ☐ = 100
25 + ☐ = 100	62 + ☐ = 100

(b)

100 − ☐ = 11	100 − ☐ = 99
100 − ☐ = 21	100 − ☐ = 89
100 − ☐ = 31	100 − ☐ = 79
100 − ☐ = 41	100 − ☐ = 69

6.3 Representing numbers up to 100 (1)

 Learning objective Recognise the place value of 2-digit numbers

 Basic questions

1 Count the dots and then write the numbers.

(a)

Tens	Ones
●●	●●●
●●	●●●

Tens place	Ones place

Written as: [ ]

(b)

Tens	Ones
●●●	●
●●	●

Tens place	Ones place

Written as: [ ]

(c)

Tens	Ones
●●●●	
●●●●	

Tens place	Ones place

Written as: [ ]

2 Fill in the blanks and draw the dots. The first one has been done for you.

(a)

Tens place	Ones place
3	5

Tens	Ones
⚪ ⚪ ⚪	⚫ ⚫ ⚫ ⚫ ⚫

Written as: 3 5

(b)

Tens place	Ones place

Tens	Ones

Written as: 6 8

(c)

Tens place	Ones place

Tens	Ones

Written as: 5 4

(d)

Tens place	Ones place

Tens	Ones

Written as: 7 0

3 Draw a line to match.

(a)

| Ones place: 2 |
| Tens place: 7 |

| 2 | 7 |

| 7 | 2 |

(b)

| 5 tens and 9 ones |

| 5 | 9 |

| 9 | 5 |

(c)

| 3   8 |

| 8 tens and 3 ones |

| 3 tens and 8 ones |

Challenge and extension questions

4 Fill in the blanks.

(a) 9 tens and 5 ones make ⬚. 6 tens and 1 one make ⬚.

(b) 47 is made up of ⬚ tens and ⬚ ones.

52 is made up of ⬚ tens and ⬚ ones.

(c) In a number 84, "4" is in the _____ place. It means 4 _____.

"8" is in the _____ place. It means 8 _____.

(d) In a number 37, "7" is in the _____ place. It means 7 _____.

"3" is in the _____ place. It means 3 _____.

(e) Some 2-digit numbers are all less than 54 and have 5 in the tens places digit. They are _____.

5 Use the numbers 6, 3, 0 and 2 to form 2-digit numbers.

They are _____.

6.4 Representing numbers up to 100 (2)

Learning objective Recognise and compare numbers to 100

Basic questions

1 Answer the questions using the number lines.

(a) Mark 8, 26, 41, 57, 74, 88 and 92 on the number line.

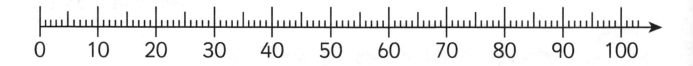

(b) Write the numbers that a, b, c, d, e, f, g and h represent.

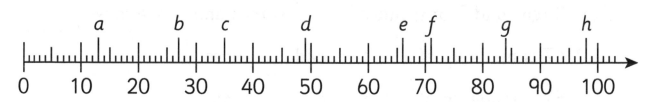

a = ☐   b = ☐   c = ☐   d = ☐

e = ☐   f = ☐   g = ☐   h = ☐

(c) Write the numbers before and after a, b, c, d, e, f, g and h.

| | a | | | | b | | | | c | | | | d | |
| | e | | | | f | | | | g | | | | h | |

(d) Write the tens numbers before and after a, b, c, d, e, f, g and h.

	a				b					c					d	
	e				f					g					h	

2 Draw lines to match the numbers to the correct sets.

15			5
49		7 in the tens place	27
70			35
72		5 in the ones place	65
78			77
90			79
93		Greater than 80	81
100			95

3 Write all the 2-digit numbers based on the information given.

(a) The digit in the ones place is 3:

_____

(b) The digit in the tens place is 6:

_____

(c) The digit in the tens place is 5:

_____

(d) It is greater than 48 and has the same digit in the tens place and ones place:

_____

(e) It has 7 in its tens place. The digit in its ones place is 7 less than the digit in its tens place:

_____

Challenge and extension question

4 Use the information to find the 2-digit numbers.

(a) The digit in the ones place is 2 greater than that in the tens place. Write all such 2-digit numbers.

_____

(b) The digit in the tens place is 2 greater than that in the ones place. Write all such numbers.

_____

6.5 Comparing numbers within 100 (1)

 Learning objective Compare and order numbers to 100

 Basic questions

1 Mark 18, 23, 32, 55, 68, 71 and 97 on the number line.

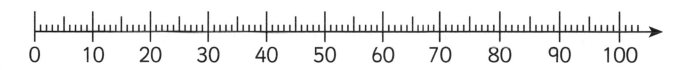

2 Write the numbers indicated.

(a) Write the numbers that a, b, c, d, e, f, g and h represent.

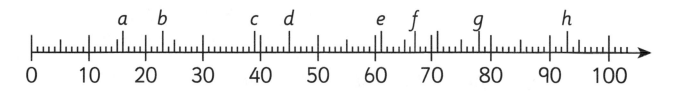

a = ☐   b = ☐   c = ☐   d = ☐

e = ☐   f = ☐   g = ☐   h = ☐

(b) Write the tens numbers that are before and after a, b, c, d, e, f, g and h.

	a				b				c				d	
	e				f				g				h	

3 Find patterns and fill in the boxes with suitable numbers.

26		28		30			33

58		62			

	85			79	77		

		71		

4 Write >, < or = in each ◯.

20 ◯ 30   13 ◯ 31   29 ◯ 30   100 ◯ 90

45 ◯ 55   27 ◯ 21   69 ◯ 50   54 ◯ 45

62 ◯ 58   0 ◯ 100   70 ◯ 71   66 ◯ 63

89 ◯ 90   46 ◯ 64   39 ◯ 40   100 ◯ 50 + 50

5 Put the numbers in order from the least to the greatest.

(a) 27   19   91   74   58   93

◯ < ◯ < ◯ < ◯ < ◯ < ◯

(b) 6   46   76   86   60   96

◯ < ◯ < ◯ < ◯ < ◯ < ◯

6 Write the following numbers in the ▢ as indicated.

90   15   28   39   47   56   34   71

The numbers greater than 45 are ▢▢▢▢.

The numbers less than 45 are ▢▢▢▢.

7 It is a 2-digit number.

The difference between the digits in the ones place and the tens place is 3.

Find all the numbers that fit the clues.

6.6 Comparing numbers within 100 (2)

Learning objective Position, compare and order numbers to 100

Basic questions

1 Fill in the boxes.

4 tens equal ☐.

6 tens equal ☐.

2 tens equal ☐.

There are ☐ tens and ☐ ones in 56.

There are ☐ tens and ☐ ones in 82.

☐ tens and ☐ ones equals 36.

☐ tens and ☐ ones equals 98.

2 (a) Write the number that comes before each number.

☐ 30   ☐ 67   ☐ 79   ☐ 90   ☐ 51

(b) Write the number that comes after each number.

40 ☐   59 ☐   67 ☐   84 ☐   99 ☐

(c) Write the numbers that come before and after each number.

☐ 63 ☐   ☐ 50 ☐   ☐ 61 ☐   ☐ 48 ☐

(d) Write the tens numbers that come before and after each number.

☐ 26 ☐   ☐ 58 ☐   ☐ 77 ☐   ☐ 80 ☐

3 Fill in the ☐ and mark the numbers on the number line.

(a) The number that comes before 84 is ☐. The third number after 84 is ☐.

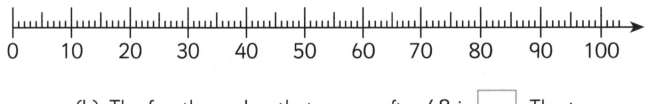

(b) The fourth number that comes after 68 is ☐. The tens number that comes before 94 is ☐.

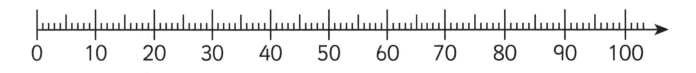

4 Write >, < or = in each ◯.

47 ◯ 74   100 ◯ 10   48 ◯ 4 tens and 8 ones

33 ◯ 31   64 ◯ 34   86 ◯ 8 ones and 6 tens

72 ◯ 76   89 ◯ 98   10 ◯ 10 ones

17 ◯ 77   30 ◯ 60   64 ◯ 46 ones

5 Put the numbers in order from the greatest to the least.

(a) 66   86   17   99   27   88

◯ > ◯ > ◯ > ◯ > ◯ > ◯

(b) 100   26   44   39   71   65

◯ > ◯ > ◯ > ◯ > ◯ > ◯

Challenge and extension question

6 Sort these numbers:

18, 80, 8, 78, 38, 82, 89, 88, 58 and 28.

Write them in the boxes below.

8 in the ones place   8 in the tens place

Greater than 50   Less than 50

6.7 Practice and exercise (I)

 Learning objective Compare and partition 2-digit numbers

 Basic questions

1 Mark 38, 42, 59, 81, 9, 76 and 95 on the number line.

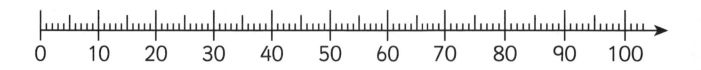

0 10 20 30 40 50 60 70 80 90 100

2 Write >, < or = in each ◯.

39 ◯ 30   41 ◯ 31   73 ◯ 49   51 ◯ 89

80 ◯ 79   6 ◯ 66   65 ◯ 38   10 ◯ 24

3 Find the nearest numbers by adding 1 or subtracting 1.

16 − 1 = ☐   29 − 1 = ☐   87 − 1 = ☐

16 + 1 = ☐   29 + 1 = ☐   87 + 1 = ☐

60 − 1 = ☐   99 − 1 = ☐   50 − 1 = ☐

60 + 1 = ☐   99 + 1 = ☐   50 + 1 = ☐

4 Write the numbers that come before and after each number.

☐ 27 ☐   ☐ 49 ☐   ☐ 65 ☐   ☐ 43 ☐

☐ 81 ☐   ☐ 30 ☐   ☐ 77 ☐   ☐ 62 ☐

5 Write the tens numbers that come before and after each number.

☐ 39 ☐   ☐ 78 ☐   ☐ 51 ☐   ☐ 62 ☐

☐ 40 ☐   ☐ 80 ☐   ☐ 65 ☐   ☐ 98 ☐

6 Get back to the tens.

31 − ☐ = 30   89 − ☐ = 80   26 − ☐ = 20

32 − ☐ = 30   77 − ☐ = 70   49 − ☐ = 40

33 − ☐ = 30   65 − ☐ = 60   64 − ☐ = 60

34 − ☐ = 30   53 − ☐ = 50   87 − ☐ = 80

7 Add up to the tens.

1 + ☐ = 10   19 + ☐ = 20   22 + ☐ = 30

31 + ☐ = 40   28 + ☐ = 30   57 + ☐ = 60

51 + ☐ = 60   37 + ☐ = 40   63 + ☐ = 70

71 + ☐ = 80   46 + ☐ = 50   76 + ☐ = 80

8 Calculate with reasoning.

16 + 4 = ☐   18 + 2 = ☐   7 + 23 = ☐

26 + 4 = ☐   38 + 2 = ☐   7 + 43 = ☐

36 + 4 = ☐   58 + 2 = ☐   7 + 63 = ☐

46 + 4 = ☐   78 + 2 = ☐   7 + 83 = ☐

Challenge and extension question

9 Make up to the tens.

58 + ☐ = ☐   65 − ☐ = ☐

74 + ☐ = ☐   87 − ☐ = ☐

☐ + 41 = ☐   ☐ − 4 = ☐

☐ + 62 = ☐   ☐ − 9 = ☐

6.8 Knowing money (1)

Learning objective Recognise, describe and compare quantities of money

Basic questions

1 Write the value of each coin.

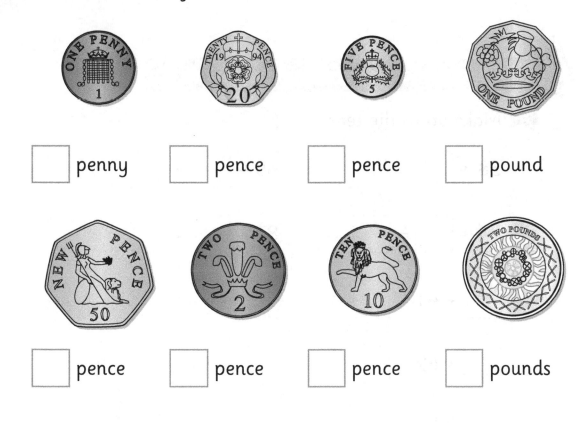

☐ penny ☐ pence ☐ pence ☐ pound

☐ pence ☐ pence ☐ pence ☐ pounds

2 Put the values of the coins in Question 1 in order, starting from the greatest.

£☐ £☐ ☐ p ☐ p ☐ p ☐ p ☐ p ☐ p

3 Write the values of the notes.

 pounds

 pounds

 pounds

 pounds

Put the values of the notes in order, starting from the least.

£ ☐   £ ☐   £ ☐   £ ☐

4 Fill in the blanks. One has been done for you.

(a) One = <u>Two</u>   (b) One = ____

(c) One = ____   (d) One = ____

(e) One = ____   (f) One = ____

5 Anya paid exactly 10 pounds for a dinosaur book.

(a) If she used only one note to pay, the value of the note is

[   ].

(b) If she used two notes to pay, the value of each note is

[   ].

(c) If she used one note and three coins to pay, the values of the note and the coins are

_____.

Challenge and extension question

6 Can you make up 18 pence using four coins? How about five coins?

Use number sentences to show all the possible ways. You can use the same coin more than once.

Four coins: _____

Five coins: _____

6.9 Knowing money (2)

 Learning objective Recognise, describe and compare quantities of money

 **Basic questions**

1 Write the total amount of money in each part.

(a)

[ ] pounds

(b)

[ ] pounds [ ] pence

(c)

[ ] pounds [ ] pence

(d)

[ ] pounds [ ] pence

2 Write the answer in each box.

(a) One 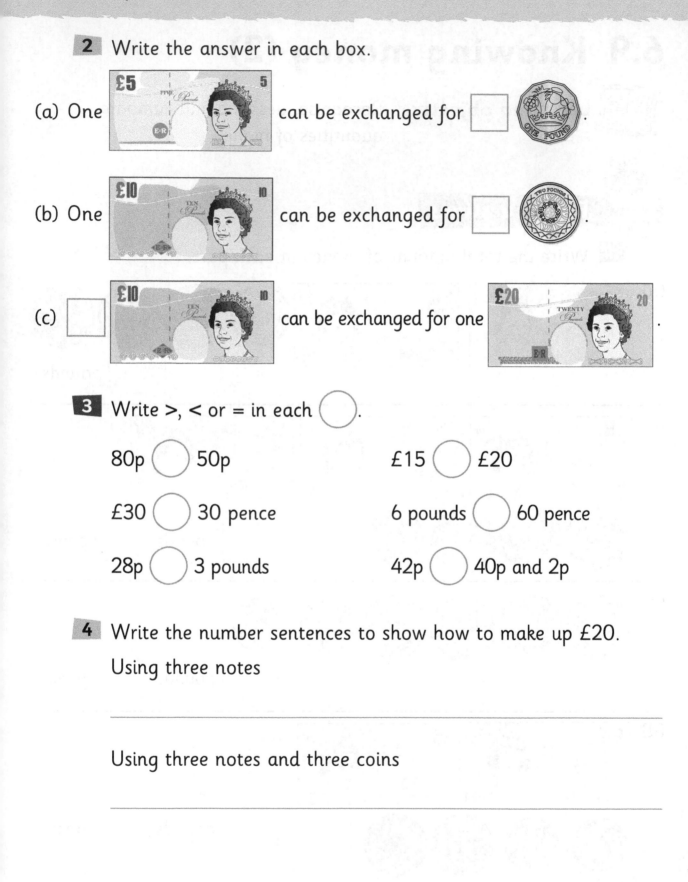 can be exchanged for [   ] .

(b) One £10 can be exchanged for [   ] .

(c) [   ] £10 can be exchanged for one £20 .

3 Write >, < or = in each ◯ .

80p ◯ 50p   £15 ◯ £20

£30 ◯ 30 pence   6 pounds ◯ 60 pence

28p ◯ 3 pounds   42p ◯ 40p and 2p

4 Write the number sentences to show how to make up £20.

Using three notes

_____

Using three notes and three coins

_____

Challenge and extension question

5 Write the number sentences to solve word problems.

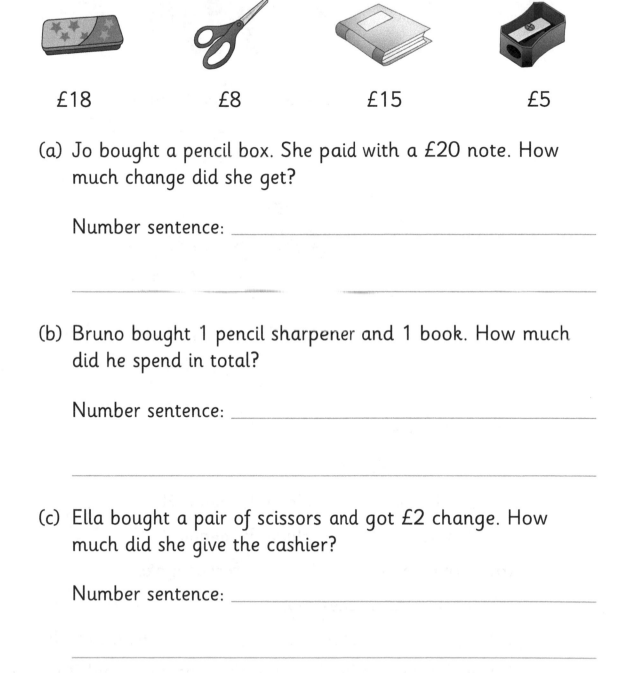

£18   £8   £15   £5

(a) Jo bought a pencil box. She paid with a £20 note. How much change did she get?

Number sentence: _____

_____

(b) Bruno bought 1 pencil sharpener and 1 book. How much did he spend in total?

Number sentence: _____

_____

(c) Ella bought a pair of scissors and got £2 change. How much did she give the cashier?

Number sentence: _____

_____

Chapter 6 test

1 Count and then fill in the missing numbers.

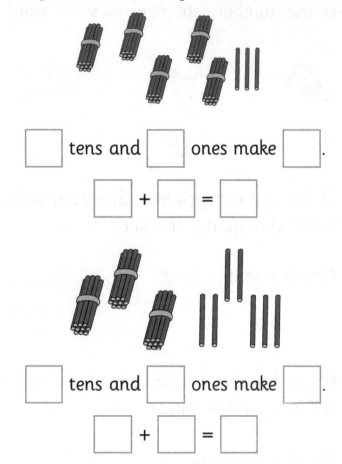

☐ tens and ☐ ones make ☐.

☐ + ☐ = ☐

☐ tens and ☐ ones make ☐.

☐ + ☐ = ☐

2 Fill in the spaces.

(a) In the number 82,

8 is in the _____ place. It means 8 _____.

2 is in the _____ place. It means 2 _____.

(b) 5 ones make _____. 5 tens make _____.

(c) 4 tens and 7 ones make _____.

(d) _____ ones and _____ tens make 38.

(e) A 2-digit number has 9 in the tens place and 2 in the ones place.

This 2-digit number is _____.

3 Answer the questions based on the number line.

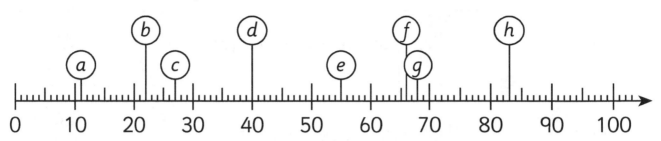

(a) Write the numbers that a, b, c, d, e, f, g and h represent.

a = ☐   b = ☐   c = ☐   d = ☐

e = ☐   f = ☐   g = ☐   h = ☐

(b) Write the numbers that come before and after a, b and c.

☐ | a | ☐   ☐ | b | ☐   ☐ | c | ☐

(c) Write the tens numbers that come before and after d, e and f.

☐ | d | ☐   ☐ | e | ☐   ☐ | f | ☐

4 Fill in the ☐ with suitable numbers.

(a)  ☐ pounds

(b)  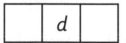 ☐ pounds

(c)    ☐ pounds ☐ pence

5 Exchange the same amount of money.

(a) One [£50 note] can be exchanged for [ ] [£10 note] .

(b) One [£20 note] can be exchanged for [ ] [£10 note] .

(c) One [£10 note] can be exchanged for [ ] [£2 coin] .

6 Solve the word problems.

[pen] Each costs £15

[ruler] Each costs £2

[storybook - Stories] Each costs £9

[eraser] Each costs £1

(a) Lily bought a pen. She paid with a £20 note. How much change did she get?

[ ] ◯ [ ] = [ ]

Lily got _____ change.

(b) Imran bought a storybook. He got 1 pound change. How much did he give to the cashier?

[ ] ◯ [ ] = [ ]

Imran gave _____ to the cashier.

(c) If you had 20 pounds, which three items would you buy?

I would like to buy _____,

_____ and _____.

[ ] ◯ [ ] ◯ [ ] = [ ]

Chapter 7　Introduction to time (I)

7.1 Year, month and day

Learning objective Recognise and use language relating to days, weeks, months and years

Basic questions

1 Write the missing word in each space.

(a) There are _____ months in a year.

The months are January, February, March, _____,
May, June, _____, August,
September, _____, November and _____.

This month is _____ .

(b) There are _____ days in a week.

The days of the week are Sunday, _____, Tuesday,
Wednesday, _____, Friday and _____ .

Today is _____ .

2 Write the following in order starting with winter.

| spring | autumn | winter | summer |

3 A festival starts on Monday 14th May 2018 and lasts 4 days.

(a) Which date is the last day of the festival?

Write the date in words and numbers:

_____

Write the date in numbers only:

_____

(b) Did the event take place on Friday of the same week?

_____ (Write yes or no.)

4 Complete the table to show public holidays in 2018.

	Dates in words and numbers	Dates in numbers	Day of the week
New Year's Day	1st January 2018		Monday
Good Friday		30/03/2018	
Christmas Day	25th December 2018		Tuesday
Boxing Day		26/12/2018	

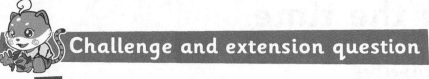

5 On which day did your school's most recent holidays start, and on which day did they end?

(a) Write the dates in words and numbers and in numbers only.

Start date:

In words and numbers: _____ .

In numbers: _____

End date:

In words and numbers: _____ .

In numbers: _____

(b) What days of the week did they fall on?

Start date: _____

End date: _____

(c) How many days did the holidays last?

_____ days

7.2 Telling the time

Learning objective

Read the time to the hour and half past the hour

Basic questions

1 Put the following days in order starting from the earliest.

(a) tomorrow, yesterday, today

_____   _____   _____

(b) evening, afternoon, morning, noon

_____   _____

_____   _____

2 What can you complete in a minute and in an hour?
Put a tick (✓) for each activity. One has been done for you.

In a minute   In an hour

(a)   Writing 10 words   ✓   ☐

(b)   Reading a story book   ☐   ☐

(c) Drinking a cup of water 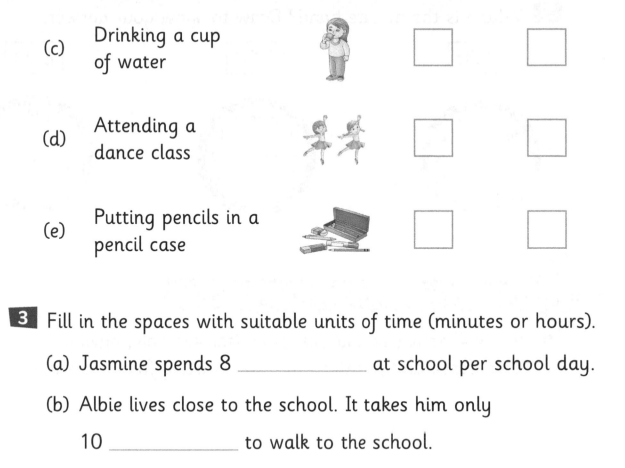 ☐ ☐

(d) Attending a dance class ☐ ☐

(e) Putting pencils in a pencil case ☐ ☐

3 Fill in the spaces with suitable units of time (minutes or hours).

(a) Jasmine spends 8 _____ at school per school day.

(b) Albie lives close to the school. It takes him only

10 _____ to walk to the school.

(c) Theo's music lesson lasted 1 _____ .

(d) Bella took 30 _____ to finish her lunch.

4 Draw a line to match the time. One has been done for you.

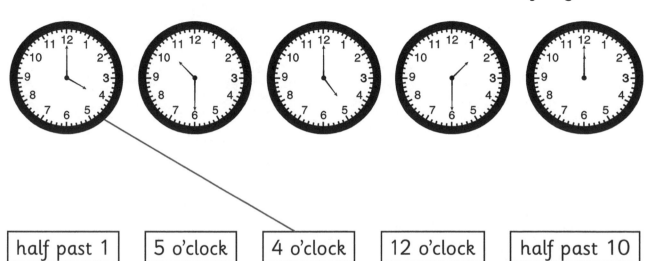

| half past 1 | 5 o'clock | 4 o'clock | 12 o'clock | half past 10 |

5 Where is the minute hand? Draw to show your answer.

| 2:00 | 9:00 | 6:00 |

Challenge and extension question

6 Write 4 activities you did today (for example, having breakfast), and put them in order starting from the earliest.

_____

_____

_____

_____

7.3 Hour and half an hour

1 Read the time on each clock. Circle the box that shows the right time.

6 o'clock	half past 6

9 o'clock	half past 9

half past 11	11 o'clock

half past 2	2 o'clock

2 Draw a line to match.

10:00	07:30	06:00	03:30

3 Put the times in order starting from the earliest.

| 3 o'clock afternoon | half past 5 morning | half past 9 evening |

| 1 o'clock afternoon | 12 noon | 5 o'clock afternoon |

_____

_____

_____

_____

_____

_____

4 Read the time and then write digital times in the spaces.

(a)   The swimming lesson starts at .   _____

    It ends at .   _____

(b)   We start our walk at .   _____

    and get back home at .   _____

5 Draw the hour hand and the minute hand.

4 o'clock

half past 9

half past 10

half past 7

Challenge and extension question

6 Which of the clocks does not work properly? Write "**X**" to indicate your answer.

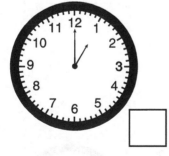

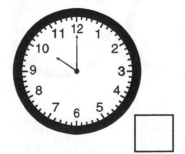

Chapter 7 test

1 Put the following months in order starting from the earliest month in the year:

September, June, January, December, August, April

_____

_____

_____

_____

_____

_____

2 If today is Monday, then

yesterday was _____,

the day before yesterday was _____,

tomorrow will be _____

and the day after tomorrow will be _____.

3 Draw a line to match.

4 Write the times in order, starting with the earliest.

| 3 o'clock afternoon | | 8 o'clock morning | | half past 11 morning |
| half past 9 evening | | 4 o'clock afternoon | | 12 noon |

_____

_____

_____

_____

_____

5 Complete the table to show the public holidays in 2017.

	Dates in words and numbers	Dates in numbers	Day of the week
New Year's Day			Sunday
Good Friday	14 April 2017		
Spring Bank Holiday		29/05/2017	Monday
Christmas Day			Monday
Boxing Day	26 December 2017		

6 Complete each statement with the correct unit of time.

hours   minutes   days

(a) A lesson lasts 35 _____ .

(b) A movie lasts 2 _____ .

(c) Ben walks from home to school in 15

_____ .

(d) There are 14 _____ in two weeks.

(e) On Saturday, Amy and her parents spent 5

_____ visiting a theme park.

7 Draw the hour hand and minute hand on each clock face to show the given time.

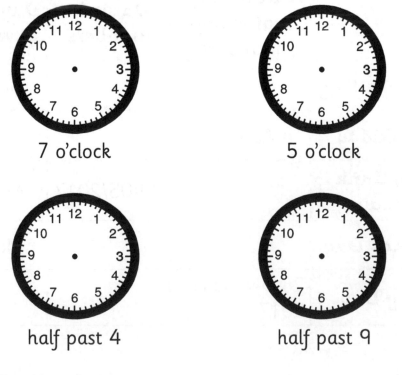

7 o'clock   5 o'clock

half past 4   half past 9

8 Write the correct time underneath each clock.

Morning break: _____

Arriving in school: _____

Lunch: _____

Breakfast: _____

Getting up: _____

Attending class: _____

Bedtime: _____

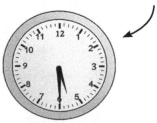

Sports activity: _____

Watching TV: _____

Dinner: _____

Chapter 8   Let's practise geometry

8.1 Left and right (1)

Learning objective Recognise and use left and right to describe position and direction

Basic questions

1 Who can do it quickly?

(a) Put out your right hand.

(b) Lift your left foot.

(c) Touch your left ear.

(d) Blink your right eye.

(e) Touch your left ear with your left hand.

(f) Pat your left foot with your right hand.

2 Fill in the blanks.

(a) I live on the left of house number 8.

I live in house number _____.

(b)

(i) ![cat] is on the _____ of ![dog]

and on the _____ of ![rabbit] .

(ii) ![rabbit] is on the _____ of ![frog]

and on the _____ of ![cat] .

(c) The children coming downstairs

are on the _____ side.

The child going upstairs is on the

_____ side.

(d) The cabbages are on the _____

side and the carrots are on the _____ side.

(e)

(i) Counting from the left, the banana is in the

_____ place.

On the left of the banana is the _____ .

(ii) There are _____ pieces of fruit on the left of the pear.

The grapes are on the _____ .

(iii) There are _____ types of fruit altogether.

The first fruit from the left is _____ .

Challenge and extension question

3 Fill in the spaces with letters to stand for the animals.

A   B   C   D   E   F

(a) On the left of the zebra are _____ , and on the

right are _____ .

(b) The kangaroo is on the _____ of the zebra.

The duck is on the _____ of the rabbit.

(c) There are _____ animals on the right of

the rabbit and there are _____ animals on

the left.

(d) _____ is the sixth from the right and the sheep

is the _____ from the left.

8.2 Left and right (2)

Learning objective Recognise and use left and right to describe position and direction

Basic questions

1 Read the instructions below and then draw a line to put each toy in its correct place on the shelves.

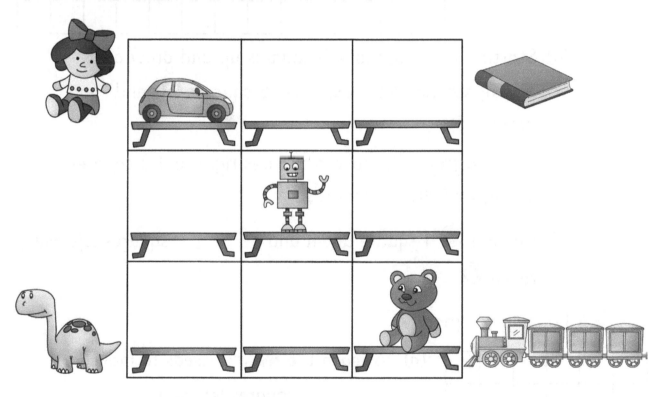

(a) On the left of the robot is a dinosaur.

The book is on the right of the robot.

(b) The doll is on the left of the teddy.

(c) The train is on the right of the car.

2 Look at the grid and answer the questions below.

Up

Left ←✛→ Right

Down

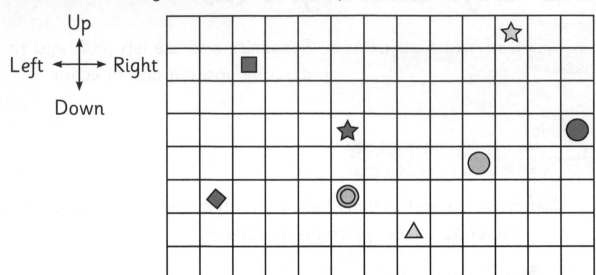

(a) Starting from △, move 3 squares up and draw a □.
Again, starting from △, move 5 squares left and draw a ◇.

(b) Starting from ☆, move 5 squares right and 3 squares down, and then draw a △.

(c) Moving ◯ 1 square down and _____ squares left will reach ◎.

3 Looking for food.

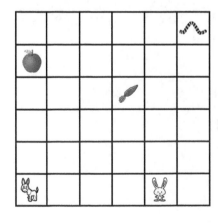

(a) To reach the 🍎, ～ needs to move _____ squares left and _____ square down.

(b) To get the 🥕, 🐰 needs to move _____ square to the _____ and then _____ squares up.

(c) Use ——→ to show how 🐰 should go to get 🥕.

Challenge and extension question

4 Draw a line to match the animals to their homes.

Can you help the animals find their homes?

On the right of my home is Kitty's.

Kitty lives on the left of my home.

 1  2  3  4

On the left of frog's home is mine.

I am Kitty. Where is my home?

8.3 Left, centre and right, top, middle and bottom

Learning objective Solve position and movement problems on grids

Basic questions

1 Guess who lives on each floor.

Asif, Marlon and Cleo decorated their balconies with flowers.

Asif's flat is below Marlon's.

Marlon's flat is below Cleo's.

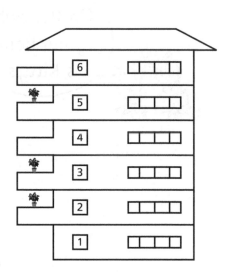

Asif lives on the _____ floor.

Marlon lives on the _____ floor.

Cleo lives on the _____ floor.

2 The toys are in their toy shelf spaces.

Which toys are:

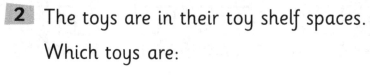

(a) on the middle shelf? _____

(b) on the top shelf? _____

(c) on the bottom shelf? _____

(d) next to the zebra's space? _____

(e) above the hen's space? _____

(f) just below the rabbit's space? _____

3 Follow the instructions to fill each grid with numbers.

(a) 10 is in the centre of the grid.

9 is above 10.

2 is on the right of 9.

1 is below 10.

8 is on the left of 1.

3 is on the left of 10.

4 is above 3.

5 is on the right of 10.

7 is below 5.

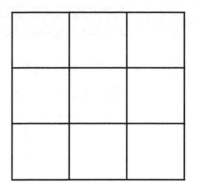

(b) 5 is in the centre of the grid.

2 is on the right of 8.

7 is below 5.

10 is on the left of 7.

8 is above 5.

3 is on the left of 5.

4 is above 3.

5 is on the left of 1.

6 is on the right of 7.

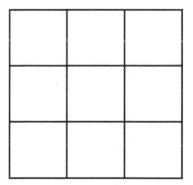

Challenge and extension question

4 Look at the grid and write directions or routes to the destinations. You cannot go through the ▓.

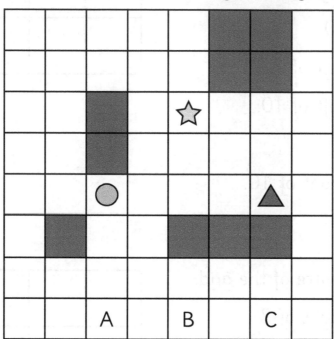

(a) Starting from C

Move 1 square right.

Move 3 squares up.

Move 5 squares left.

Reach _____

(b) Starting from B

_____

_____

_____

Reach ▲

(c) Starting from A

_____

_____

_____

Reach ☆

8.4 Comparing lengths

 Learning objective Describe and compare lengths of objects

 Basic questions

1 Which ribbon is longer? Put a ✔ in the box.

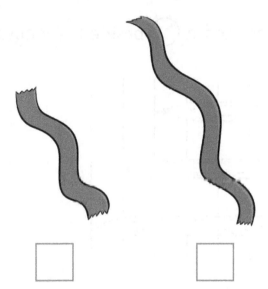

2 Compare the heights of the animals. Put a ◯ in the box below the tallest and a △ in the box below the shortest.

3 Put a ✔ beside the shortest pencil and a ◯ beside the longest.

4 Put a ✔ beside the shortest and a ◯ beside the longest.

5 Put a ✔ beside the shortest rope and a ◯ beside the longest.

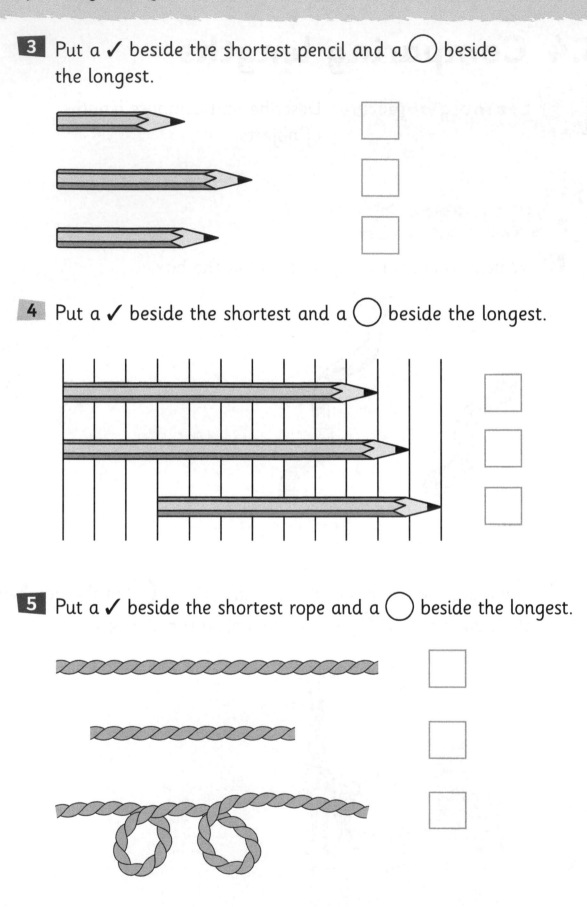

Challenge and extension question

6 Which monkey has the longer way to reach the bananas?
Put a ✓ in the box.

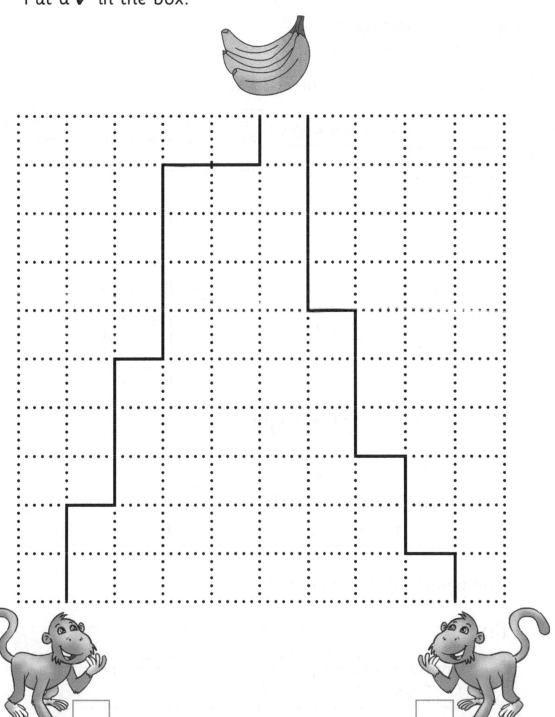

8.5 Length and height (1)

Learning objective Measure and record lengths

Basic questions

1 Put a ✓ next to the correct way to measure.

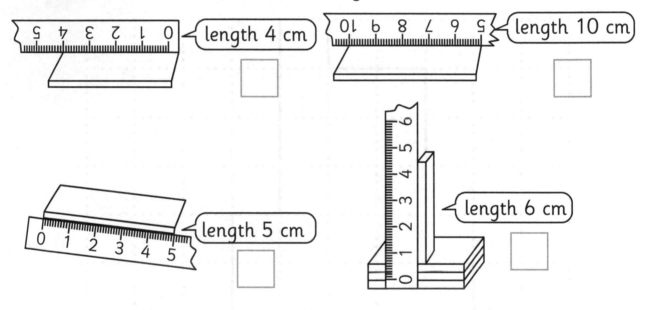

length 4 cm

length 10 cm

length 5 cm

length 6 cm

2 Take a measure of the length.

(a)

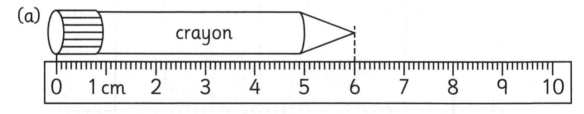

crayon

The length of the crayon is ☐ cm.

(b)

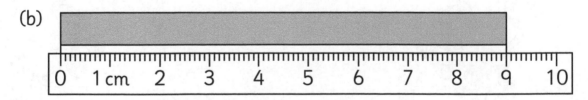

The length of the paper strip is ☐ cm.

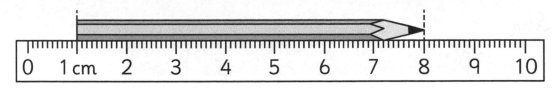

The pencil is ☐ cm long.

3 Use a ruler to measure these objects in your classroom.

(a)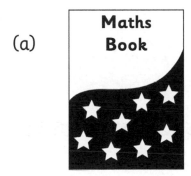

The width of the book is ☐ cm.

(b)

The length of the pencil is ☐ cm.

(c)

The length of the pencil case is ☐ cm.

Challenge and extension question

4 Fill in the ☐ with your estimate of the length of each piece of string. Check your estimate using a ruler.

estimate ☐ cm   actual ☐ cm

estimate ☐ cm   actual ☐ cm

estimate ☐ cm   actual ☐ cm

estimate ☐ cm   actual ☐ cm

estimate ☐ cm   actual ☐ cm

estimate ☐ cm   actual ☐ cm

estimate ☐ cm   actual ☐ cm

8.6 Length and height (2)

Learning objective Measure and record lengths

Basic questions

1 Use a ruler to measure the lengths and fill in the boxes.

(a) The rope is ☐ cm long. If 4 cm is cut off, it will be ☐ cm long.

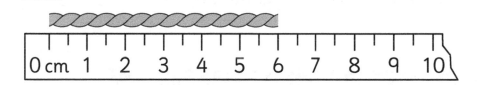

(b) The snail has covered ☐ cm. It has ☐ cm to reach the end.

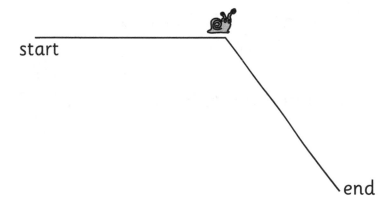

start

end

2 Write the missing number in each ☐.

9 m   7 m

☐ m

40 cm   ☐ cm

70 cm

3 Fill in the spaces with suitable units of length.

James is 98 _____ tall.

The height of a door is 2 _____.

The length of the shoe is 18 _____.

The length of the rubber is 3 _____.

The skipping rope is 3 _____ long.

The length of the playground is 100 _____.

Challenge and extension question

4 Application problems.

(a) 🧒 is 90 cm tall. 🧒 is 98 cm tall.

How many centimetres taller is 🧒 than 🧒 ?

Answer: _____

(b) A rope is 80 metres long.
It was cut by 8 metres to make a 〜.
How many metres are left?

Answer: _____

8.7 Practice and exercise (II)

Learning objective Describe and compare different quantities of length and time

Basic questions

1 (a) Mark 27, 49, 85 and 92 on the number line.

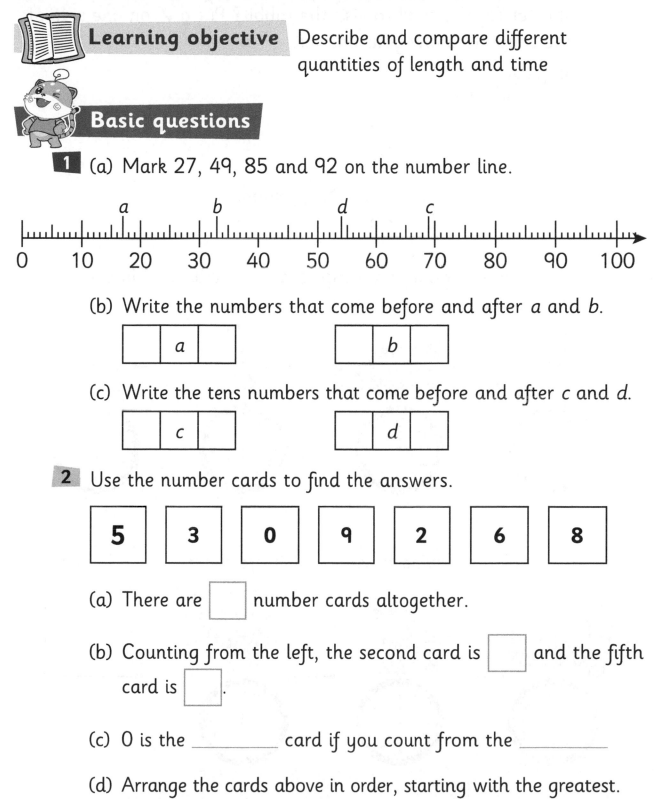

(b) Write the numbers that come before and after *a* and *b*.

	a	

	b	

(c) Write the tens numbers that come before and after *c* and *d*.

	c	

	d	

2 Use the number cards to find the answers.

5	3	0	9	2	6	8

(a) There are ☐ number cards altogether.

(b) Counting from the left, the second card is ☐ and the fifth card is ☐.

(c) 0 is the _____ card if you count from the _____

(d) Arrange the cards above in order, starting with the greatest.

3 Measure the length and then fill in the ☐. Which way is shorter for the snail to visit the rabbit? Put a ✓ on one side of the shape to show your answer.

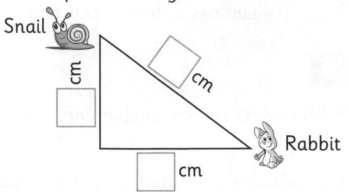

Snail

cm

cm

Rabbit

cm

4 The school is going on a trip to a zoo. Write the time in each box.

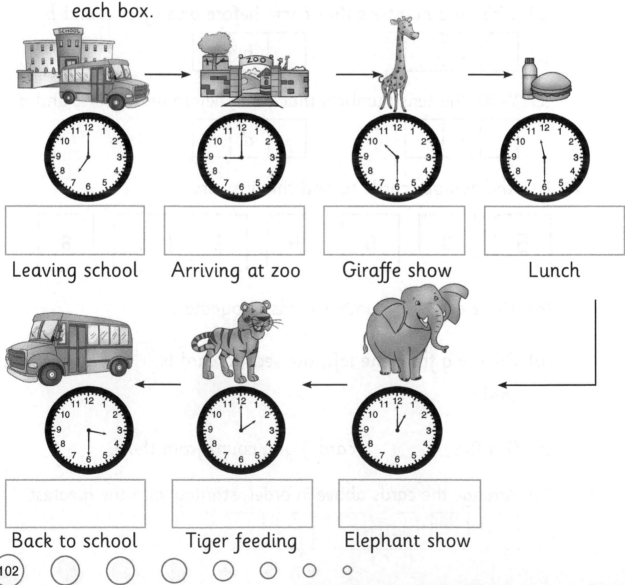

Leaving school   Arriving at zoo   Giraffe show   Lunch

Back to school   Tiger feeding   Elephant show

5 Write the correct answer in each box.

(a) The digit in the tens place of 47 is ☐.

It means 4 _____

(b) 3 ones and 7 tens make ☐.

It is ☐ greater than 72.

(c) The tens nearest to 68 is ☐.

The difference between these two numbers is ☐.

(d) A number has 6 in its tens place, and the digit in the ones place is 2 greater than that in the tens place. The number is ☐.

(e) If you arrange the numbers 20, 42, 6, 89, 100 and 12 in order, from the least to the greatest, the fourth number will be ☐.

Challenge and extension question

6 Snail and ant are having a race.

60 cm   70 cm   80 cm   90 cm   100 cm

(a) The ant has covered ☐ cm and it is ☐ cm away from the finish line.

(b) The snail has covered ☐ cm and it has ☐ cm left to reach the finish line.

(c) _____ is likely to reach the finish line first.

Chapter 8 test

1 Use a ruler to measure each length and then fill in the boxes.

(a)

☐ cm   ☐ cm

☐ cm

(b) The little snail is going home.

> I have moved ＿＿ cm and still have ＿＿ cm left.

2 Choose a suitable unit, putting a ✓ in the ◯ to show which one you would use.

The height of the camel is about

| 2 m ◯ | 2 cm ◯ |

The length of the blackboard is about

| 3 m ◯ | 3 cm ◯ |

The length of the pen is about

| 15 m ◯ | 15 cm ◯ |

The height of the lamp is about

| 30 m ◯ | 30 cm ◯ |

3 Compare the lengths. Colour the longest pencil in red.
Colour the shortest pencil in yellow.

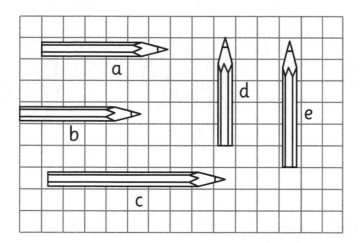

4 Use letters to indicate the animals for each question below.

| | Left | Centre | Right |

Left   Centre   Right

(a) In the picture showing , the animal living in the ■

is ____.

A    B 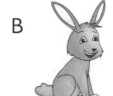   C

(b) The animals living next to [tiger] are [lion] and _____.

A [monkey] B [elephant] C [cow]

(c) Which animal is left of [camel] ? _____.

A [rabbit] B [elephant] C [lion]

(d) Which animal is right of [monkey] ? _____.

A [elephant] B [panda] C [tiger]

End of year test

1 Work these out mentally. Write the answers. (10%)

11 – 4 = ☐   9 + 9 = ☐   2 + 6 = ☐   9 – 2 = ☐

6 – 5 = ☐   7 – 4 = ☐   6 + 4 = ☐   10 – 6 = ☐

10 + 10 = ☐   3 – 0 = ☐

2 Write the numbers that come before and after. (2%)

	39	

3 Find a pattern and then fill in with numbers. (2%)

	20	30		50	

4 Write >, < or = in each ◯ and write a number in each ☐. (16%)

25 ◯ 28   75 ◯ 75 – 1   0 + 5 ◯ 10 – 5

☐ > 96   10 – ☐ < 4   7 + ☐ > 11 – 3

88 pence ◯ 1 pound   50 pence ◯ 5 pounds

5 Find patterns and then fill in the boxes with numbers. (4%)

$8 + 8 = 16$

$9 + 7 = 16$

$10 + \boxed{} = 16$

$\boxed{} + \boxed{} = 16$

$20 - 10 = 10$

$20 - 8 = 12$

$20 - 6 = \boxed{}$

$20 - \boxed{} = \boxed{}$

6 Complete the addition table. (3%)

	6 + 5	7 + 5
5 + 6		7 + 6
5 + 7		7 + 7

7 Fill in the number wall. (3%)

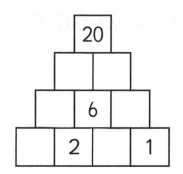

8 Use the number lines for addition and subtraction calculations. (4%)

(a)

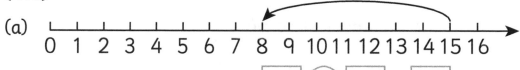

□ ○ □ = □

(b)

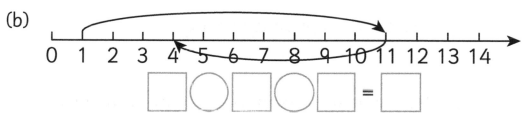

□ ○ □ ○ □ = □

9 Choose 3 numbers from

2 , **5** , **7** , **9** and **12** .

Make two addition sentences and two subtraction sentences. (8%)

Choice 1: □ □ □   Choice 2: □ □ □

10 Fill in the answers. (20%)

(a) 73 is made up of ☐ ones and ☐ tens.

10 tens are ☐ .

(b) Write 42 in words _____ .

The tens nearest to 56 is ☐ .

(c) Adding up one £5 note, two £2 coins and two 50p coins, there are ☐ pounds in total.

(d) Fill in with suitable units of length or money.

I took 10 _____ to buy a ruler and a 15 _____ long pencil in a supermarket. The cost of the ruler is 1 _____ and the cost of the pencil is 50 _____ .

(e) In a 2-digit number, the digit in the ones place is less than 1 and the digit in the tens place is greater than 8. The number is ☐ .

(f) Write the length of each shape underneath each ruler.

0 1 2 3 4 5 6

[ ] cm

3 4 5 6 7 8

[ ] cm

(g) Write the time on the clock faces in words.

_____

_____

It is _____ now.

30 minutes later, it will be _____.

(h) | 13 | | 10 | | 9 | | 1 | | 20 | | 0 | | 18 | | 16 |

Use the number cards to find the answers.

(i) The sixth number from the left is [ ].

(ii) The number closest to 10 is [ ].

(iii) Among these numbers there are [ ] numbers containing the digit 1.

(i) Ella, Harpreet, Samira and Alvin all live in a four-storey apartment building.

Ella's home is above Harpreet's.

Samira's home is below Harpreet's.

Alvin lives on the first floor.

_____ lives on the second floor.

_____ lives on the third floor.

_____ lives on the fourth floor.

11 Make a guess. (2%)

I am a teacher.

The teacher is in the middle. There are ☐ people in the queue.

12 Read the clues to find the correct object. Write the letter of each object in the box. (6%)

A ●   B ◻   C ▭

D ⬭   E △

(a) It can't roll. It has rectangular flat faces. ☐

(b) It can roll and it is not stable on the ground. ☐

(c) It can roll and it has flat and curved faces. ☐

(d) It can't roll. It has triangular flat faces. ☐

13 Work out the sums. (20%)

(a)

_____

(b) How many books are inside?

_____

There are 16 books in total.

(c)

_____

(d)   Tom went shopping with 20 pounds.

Art box: Pen: Tape measure:
12 pounds   8 pounds   5 pounds

(i) If Tom spent all his money, he could buy _____
     and _____.

(ii) If Tom bought ✏ and a 🎞, he would have
     _____ pounds left.

(iii) If Tom can at most spend half of his money, he could
     buy _____ or _____.

(iv) If Tom can spend exactly a quarter of his money, he could buy _____.

(e) After eating 6 🍎, Yee has 9 🍎 apples left. How many 🍎 did Yee have at first?

Number sentence:

_____

(f) There were 20 people on the bus. At a bus stop, 8 people got off and another 5 people got on the bus. How many people were then on the bus?

Number sentence:

_____

Notes

Notes